# MENTAL MATH

Tricks And Practical Strategies To Make
Calculations Faster, Enhance Your Math Skills And
Solve Everyday Math Problems Easily

THOMAS SCOFIELD

THOMAS SCOFIELD

Copyright © 2018 Thomas Scofield

All rights reserved.

In no way is it legal to reproduce, duplicate, or transmit any part of this document in either electronic means or in printed format. recording of this publication is strictly prohibited and any storage of this document is not allowed unless with written permission from the publisher. all rights reserved. The information provided herein is stated to be truthful and consistent, in that any liability, in terms of inattention or otherwise, by any usage or abuse of any policies, processes, or directions contained within is the solitary and utter responsibility of the recipient reader. under no circumstances will any legal responsibility or blame be held against the publisher for any reparation, damages, or monetary loss due to the information herein, either directly or indirectly. Respective authors own all copyrights not held by the publisher. The information herein is offered for informational purposes solely, and is universal as so. the presentation of the information is without contract or any type of guarantee assurance. The trademarks that are used are without any consent, and the publication of the trademark is without permission or backing by the trademark owner. all trademarks and brands within this book are for clarifying purposes only and are the owned by the owners themselves, not affiliated with this document. The author wishes to thank Freepik, Stephen Hutchings, Smashicons from www.flaticon.com for the images on the cover.

# MENTAL MATH

# TABLE OF CONTENTS

Introduction .................................................................................................7
Chapter One:
Multiplying Big Numbers Quickly................................................................9
    Multiplication by single digit numbers .................................................10
    Multiplication by powers of ten ...........................................................10
    Mental estimation ...............................................................................10
    Activities .............................................................................................11
    Multiplying by numbers around 100 ...................................................12
Chapter Two:
Proportions and Ratios ..............................................................................15
    Proportions .........................................................................................15
    Scaling examples ...............................................................................17
Chapter Three:
Trigonometry and Its Uses in Real Life .....................................................23
    Physics ...............................................................................................23
    Navigation ..........................................................................................24
    Criminology ........................................................................................24
    Marine biology ....................................................................................24
    Marine engineering ............................................................................25
    Archaeology .......................................................................................25
    Surveying ...........................................................................................25
    Trigonometric calculations .................................................................26
    SOH CAH TOA .................................................................................27
    Circles ................................................................................................27
Chapter Four:
Adding and Subtracting Fractions .............................................................29
    Solving for fractions ...........................................................................30
Chapter Five:
Mean, Median, and Standard Deviation in Everyday Life .........................37
    Findings averages in data sets..........................................................38
Chapter Six:
Working with Conversion Factors ..............................................................43
    Conversion factors .............................................................................43
Chapter Seven:
Using the PIN Technique ...........................................................................47
    The PIN method ................................................................................48
Chapter Eight:
Techniques for SAT, GMAT, and GRE Students .......................................55
    SAT problems ....................................................................................55
Chapter Nine:
Math Strategies for Anyone .......................................................................59
    Adjusted gross income ......................................................................59
    Price discounts ..................................................................................60

    Simple interest ........................................................................................ 61
    Mortgage loans ....................................................................................... 61
Conclusion ..................................................................................................... 63
Other Books By Thomas Scofield ................................................................. 64

THOMAS SCOFIELD

# Introduction

Congratulations and thank you for purchasing *Mental Math: Tricks and Practical Strategies to Make Calculations Faster, Enhance Your Math Skills, and Solve Everyday Math Problems Easily*. Mental math skills are often neglected by most. Those who do work on their mental math skills, however, are put at a great advantage in everyday life for doing so. The first step in bettering your mental math skills is to work on your basic computations. These are gone over within the first few chapters of this book until we finally start to delve into more advanced subjects near the end. Do not be intimidated by the book if you are not inclined toward mathematical thinking. The subjects and concepts gone over in this book are relatively easy and straightforward, though they will still be useful for you when dealing when more complicated real-world problems.

To that end, the following chapters will discuss a wide variety of topics including tricks to multiply big numbers mentally, how to work with proportions and ratios, trigonometry uses in real life (for example in physics, navigation, criminology, marine biology, and archeology), tricks to add and subtract fractions fast, mean median and standard deviation and mastering them in real life, tricks to work with conversion factors, usage of the Plugging In Numbers (PIN) technique in tests, techniques for SAT, GMAT and GRE students, and math strategies anyone can master.

Skills regarding these subjects can be difficult to improve upon but when some basic proficiency has been established, getting into more advanced areas of study becomes possible. This is when you may even find that you are more talented when it comes to mathematics than you may have originally thought yourself to be.

Thank you again for purchasing this book. Let's hope the information provided here helps you in your everyday life.

# Chapter One:
## Multiplying Big Numbers Quickly

Mental math skills are extremely important ones to have for use on a regular day to day basis but they require long hours of diligent practice to maintain and improve. Once basic proficiency in math has been established, it is never enough to just halt your progress wherever you decide to. Learning skills in math is just like learning skills in any other field. It is a cumulative process that requires lots of time and dedication to be done effectively.

There are many different facets of mathematics that you have to learn in order to gain any proficiency in the subject. Throughout the course of this book, we will go over some of the most important of these facets especially through the optic of developing mental math skills.

We should start with a basic aspect of math, multiplication. It takes a number of different strategies to develop mental calculation abilities in regards to multiplication. Once all of these strategies are learned, they can easily be applied to specific situations that apply to each respective strategy.

When you think of most of the math that the average person finds him or herself doing on a day to day basis, it is usually not done by means of calculators and even more rarely done by means of written computation. It is almost always done instead by mental computations. With this being said, it is no wonder why skills developed in mental math computations are always incredibly useful ones to have. Sadly though, many (if not most) neglect to develop their mental math skills beyond a certain age and are put at a disadvantage for doing so. As far as students are concerned, developing mental math skills can also help them to get a better understanding of both number sense and number properties.

## Multiplication by single digit numbers
The first type of mental multiplication that we should now go over is multiplication by single digit numbers. At first, students can use this strategy to multiply smaller numbers. Once they have developed skills with these they can then move on to bigger numbers. For example, a student can multiply 3 four times to get 12. If he or she then multiply 3 by 12 then the student gets 36.

## Multiplication by powers of ten
The next step in multiplication is multiplying by powers of ten. This is much easier to do as it only results in another decimal point being included in the final answer. To multiply by powers of ten, all you need to do is multiply the other number by the first number in the power of ten. For example, four times ten would come out to 40, while four times one would come out to four. As you can see, the decimal points, or zeros, in these cases just carry over with the rest of what is being multiplied.

If, however, the numbers do not end in zero, then the first decimal point listed should have a zero put behind it when multiplying. For example, 15 times 6 would translate to 5 times 6 (30) and 10 times 6 (60). 30 plus 60 would then equal 90, so 15 times 6 would be 90.

One strategy that students often use is the visualization of written algorithms. This is not an advisable strategy to use because it makes making computations much less efficient and much more difficult. Algorithms only work well when written down on paper. There are other much more useful methods for multiplying mentally.

## Mental estimation
Mental estimation is another very important skill for a student to learn. This is especially useful when answers are specific to an extent at which it would require a calculator to find them. For example, estimation would have to be used when finding out the area of a yard on the fly. If the fencing was 20ft. by 10ft., the area would be 200ft. A student could not, however, know definitively the length of the

fence, so he or she would have to estimate in this case. To better estimate mentally you must know how to do mental calculations, round to convenient numbers, and be able to determine whether or not an answer is too big or too small.

Again, it requires frequently repeated practices to develop skills in making mental math calculations. There are some very helpful activities for developing these skills which should now be mentioned.

## Activities

### Activity 1
Use single digit strategies to improve on your most basic multiplications. Once the more basic single digit strategies have been gone over for some time, you can then move into bigger numbers and more complex calculations, such as those by factors of ten.

### Activity 2
Apply the distributive property to the single digit strategies previously gone over. The distributive property can be solved in these three steps:

1. Multiply any term placed outside of the parentheses by each and every term located within the parentheses. In doing this you will distribute the outside term among the inside terms.
2. Combine like terms. If any terms in the equation are the same, combine them into one term.
3. Add the terms left to solve the equation.

Let's look at an example of solving an equation with the distributive property:

$$5(2+8) = 5(2) + 5(8) = 10+40 = 50, \text{ so, } 5(2+8) = 50$$

### Activity 3
Split larger numbers into factors that will fit into them and then solve the equation using these numbers. We will explore this further later.

**Activity 4**
Keep a mental track of the equations that you are solving throughout activities one and two. This will engage the circuitry involved in solving these problems long term and you will commit these strategies to your long-term memory.

**Activity 5**
Keep an eye out when multiplying large numbers for factors that allow for easy multiplication. For example, 8 times 21 can be distributed as 8 times 20 (160) plus 8 times 1 (8), so your answer here would be 168.

Now that our basic activities have been outlined, we should delve deeper into how to go about these activities in the easiest ways starting with activity number 1: multiplying by single digits.

To multiply by 2, take the number being multiplied and double it. To multiply by three, double the number and then add to that the number once more. To multiply by four, double the number two times over. To multiply by 5, multiply the number by 10, and then halve the answer. There are two main ways to go about multiplying by 6. Either you could multiply the number by 3 and then double that number, or you could multiply the number by 5 and then add the number once more. There is no easy way to multiply by 7. You must just count up by 7s by whatever you are multiplying it by. To multiply by 8, double the number three times over. To multiply by 9, first, multiply the number by 10 and then subtract the number from that answer. To multiply by 10, just add a zero to the number (but watch out for decimal places carefully).

We can then extend all those principles laid out for single digit numbers to use for dealing with larger numbers. To use the number 100 for example, there are many ways to multiply easily by 100 and numbers around it.

**Multiplying by numbers around 100**
If you wanted to multiply by 101, you could multiply by 100 and just add the number that you are multiplying once more. To multiply by a 102, first, multiply by 100 and then add the number times 2. To multiply by 99, first, multiply the number by 100 and then subtract the number being multiplied from that once. To multiply by 98, first,

multiply by 100 and then subtract the number being multiplied times 2. As you can probably see by now, many of the same rules mentioned in the first activity apply just on a grander scale.

The next helpful thing that we can do is to split the factors that we come across in equations with bigger numbers into smaller factors that can come out of them. For example, 20 times 36 is an equation with very large numbers, but we can split these numbers into multiplications that will add up to them. 20 could here become 5 times 4, and 36 here could become 6 times 6. So the result here would be 5 times 4 times 6 times 6, equaling 720, so 20 times 36 would equal 720. Another example would be 16 times 18. 16 could become 4 times 4, and 18 could become 6 times 3. So we would eventually get 4 times 4 times 6 times 3. This would equal 288, so 16 times 18 would equal 288.

Next, we should make it a point to keep track of the numbers in any given equation that you are solving, especially 0s. Multiplying by numbers which end in 0 is usually a very easy thing to do but the only stipulation is that you have to keep track of the 0s in these cases. On example of doing this could be in solving 30 times 40. This could be understood as 30 groups of 40. 30 groups of 40 could then been seen as 10 groups of three times 40. 3 groups of 40 would then be 120, and 10 groups of 120 would then be 1,200, so your final answer would be 1,200. As you can probably see, multiplying using this technique is simply not possible without paying special attention to keeping track of the 0s.

There should also be mentioned tips on multiplying by other special numbers between 1 and 100. These, like the ones near 1 and the ones near 100, can be multiplied more easily when you take into account the relations to other numbers.

Some examples of these numbers are listed below:

- To multiply a number by 50, multiply it by 100 and then halve that number.
- To multiply a number by 25, multiply it by 100 and then divide that answer by 4.

- To multiply a number by 500, multiply it by 1,000 and then divide the answer by 2.
- To multiply a number by 250, multiply it by 1,000 and then divide the answer by 4.
- To multiply a number by 125, multiply it by 1,000 and then divide the answer by 8.

Again, these tips and activities should be used and applied on a day to day basis to achieve any significant results. Mental math is just like anything else in that you need to be persistent to develop skills in it. If you do, however, stick to using the tips laid out here daily for some time, and put genuine effort into getting better at making these calculations, then you should eventually see some improvements in your performance and ability.

# Chapter Two:
## Proportions and Ratios

The next area which would be helpful to go over is proportions and ratios. We should start off by first defining these two terms. A ratio is essentially a means of comparing two quantities, whatever they might be made to represent. An example of a ratio would be miles per hour. Here, we are comparing the number of miles with the number of hours that it takes to drive those miles. A proportion, on the other hand, is an equation which assumes that the values of two or more different ratios are equivalent to one another. If the two equal one another, then we call the two being "in proportion" with each other.

We should now go a little more in-depth with our discussion of each of these. A ratio, as mentioned above, is a comparison of two terms. These are usually divided by one another when ratios are expressed. Some examples of this are as follows:

$$x \text{ to } y, \; x{:}y, \; x/y$$

**Proportions**
A proportion is another equation. The purpose of this one is to find what, if any, equations are equivalent to another. These can be extremely useful in everyday scenarios, for example, let's say that you are making cookies and the packaging says that one pack of cookie dough will make 20 cookies. A proportion that you could deduce from this is that two packs of the cookie dough would then make 40 cookies. If we were to write this out mathematically it would look something like this:

*20/1=40/2*

The writing out of a proportion requires more variables than does the writing out of a ratio because there are more terms in proportions. A proportion written out would usually be along the lines of "x is to y as z is to w", or, $x/y=z/w$.

Often times you will get ratios and or proportions in which some of the individual variables will be unknown to you. This is where some basic algebra skills will come into play as you will have to then solve these equations. If you are reading this as a non-math person then do not be intimidated. This is basic equations to follow and can be done easily by someone who does not even have a math background. Let's now look at another basic cooking example below:

You have to make 20 pancakes for a get together with family and friends. You know that making these twenty will require you to use 2 eggs. And so, 2 eggs will make a total of 20 pancakes. Using this information, how many eggs will it require to make 100 pancakes?

So, 20 pancakes=2 eggs, 100 pancakes=x eggs.

There are two common equations that could be drawn up here, but only one is correct:

*Eggs/pancakes = eggs/pancakes*
*or*
*pancakes/eggs = pancakes/eggs*

The correct equation, in this case, would be:

*eggs/pancakes=eggs/pancakes*

because this one puts out mystery number (x eggs) in the numerator, making it possible for us to solve for x.

Here is what we would do next:

1. Write it out as $x/100=2/20$
2. Then, we would multiply both sides by 100:

3. 100 times x/100, and 100 times 2/20, this would then become x=200/20, and finally, x=10.

It is, however, possible to solve for x even if the variable is in the denominator. This would just require us to use another method known as the cross product method. This one is not as commonly used but it could be argued that it is even easier than the method previously mentioned. The cross product, in this case, is the product of the first numerator of the proportion and the second denominator of it. The inverse is true for the second numerator; you would multiply it by the first denominator. Here is an example of this method using the cookie dough equation mentioned above. Again x, or 2, is represented on the bottom here:

$$20/1=40/2$$

So cross multiplying would look something like this: 1x40=2x20=40.

We should now consider again our basic equation for proportions: x/y = z/w, to cross multiply this we would come up with something like this: xy = zy.

## Scaling examples

The next topic that we should go over is scaling. This of the map keys that you have seen. They usually will show you some minuscule length like an inch and tell you that the length on the map is equal to another length, in reality, 100 miles for example. This is what is known as scaling. We often use this method to depict various sizes of things. Scaling usually consists of maintaining the proportions of a certain object, while adjusting its size in all sorts of ways. It remains the same shape, just not the same size. In scaling, you can either go up (enlarge) or go down (reduce) in the size of an object. Scaling factors are often represented by ratios, for example, 1:4 represents a quarter, 1:8 represents an eighth and so on. Thus, 1:4 of an object, in this case, would not be a quarter of the object but the object brought down to a quarter of its original size. When solving equations for the scaling of objects, it is necessary to use ratios to represent the scales. For example, if we were to reproduce a 1:4 size version of a 20ft wall, the calculation that we would have to use would look something like this:

$$20 \times 1:4 = 20 \times 1/4 = 5$$

If we do not know what we are scaling to, then x becomes our variable. For example, in a scale model of 1: x, x is our constant term, and therefore all the measures we could create would have to be 1: x- of the real measurement. The only thing left for us to do in this case would be to solve for x, which was mentioned how to do previously. The same would hold true when enlarging a figure. In depicting something in the scale of 2:1, for instance, all models to come would be twice as large as the original.

So let's move on to some word problems now. Say, for example, you are building a fence. 20 boards, in this case, could be used to build 5 feet-fence. With that being said, we now need to find out how many boards it would require to build 20 feet-fence. Let's now plug in the variables into the equation listed above for proportions, x/y=z/w.

What we are trying to solve for in this case would be z, and the equation would look something like this: 20/5 = z/20. Here, as you can see, our unknown variable is in the numerator. Our next step would look something like this:

$$20 \times 20 = 5z, \text{ so, } 400 = 5z, 400/5 = 80, \text{ and therefore, } z = 80$$

Another example in which there is only one variable to solve for is 2:x = 3:9.

The first step that we would take here would be to convert the colons into slashes to put the equation in terms of two fractions: 2/x = 3/9

The next that we would need to take here would be to cross multiply our fractions.
This would look something like this: 2 x 9 = 3x

Now, the value of 3x is equal to the value of 2 x 9 so we would have to figure out the value of the first multiplication here, 2 x 9. 2 x 9 = 18.

And so its 18 = 3x here.

Next, to solve for x we would need to divide 18 by 3. 18/3 = 6.

# MENTAL MATH

And finally, $x = 6$.

Up until this point all of the examples that we have gone over have been pretty basic, but often times you will find yourself having to solve more complicated problems with proportions, even in your everyday life. We should now up our game with problems that are a little more complicated so that we will get a better view of just how much you can do when solving for proportions and ratios.

The next example is:

$$(2x + 1):2 = (x + 2):5$$

You have probably noticed more parentheses and variables here than usual. Do not be alarmed if you are not a math person though. Equations like this are easier to solve than they may appear to be for someone who does not go over algebra regularly.

The first step here, like the first step in the previously mentioned problem, would be to convert the colons to fractions. This would make the equation look something like this:

$$2x + 1 / 2 = x + 2 / 5$$

Next, you would have to do a multiplication by the cross (keep in mind that you cross multiply the numerators with the denominators, not any terms within the parenthesis). This would make the equation then look like this:

$$5(2x + 1) = 2(x + 2)$$

And so, after multiplying the inside terms with the outside terms, on either side of this equation, what we are left with looks something like this:

$$10x + 5 = 2x + 4$$

The next step is where this gets a little tricky. Here we need to combine like terms. When this is done we need to subtract 2 from 10 and 4 from 5 in this case. The result would look like this:

$$8x = 1$$

And finally, we divide 1 by 8 to get ⅛, so x = ⅛

We should now look at a word problem. Here is our next example: 12 inches is exactly 30.48 centimeters, so, with that said, how many centimeters would there be in 30 inches? In this equation, the letter c will be used to represent centimeters. Our basic equation would then look something like this:

$$12 / 30.48 = 30 / c$$

Next, we need to cross multiply:

$$30 \times 30.48 = 12\,c$$

So, 914.4 = 12 c, and then we would divide 914.4 by 12: 914.4 / 12 = 76.2

$$c = 76.2$$

Next is another word problem. Word problems in these cases are especially useful because they give us an idea of what it is like to encounter these problems in everyday life and they prepare us for real-world problem-solving. Here is our next example: You have a metal bar that has a length of 10 feet and a weight of 128 pounds. With that being said, what would be the weight of a bar with the same density but a length of only 2 feet and 4 inches?

This is a more complicated problem than anything that we have yet come across. The first problem that we need to solve in this case is that of inch conversion, we can do this by finding the fractional form of the inches in relation to feet. Our equation for 2 feet 4 inches would look something like this:

$$2\,ft. + 4\,in. = 2\,ft. + 1/3\,ft. = 7/3\,ft.$$

Now, to add all of the other variables:

$$10 / 128 = (7/3) / w$$

Next, we would have to cross multiply:

$$(7/3) \times 128 = 10\,w$$

$$298.\,1\,/\,3 = 10\,w$$

Next we would have to divide 298. 1 / 3 by 10:

$$298.\,1\,/\,3\,/\,10 = 29.\,8\,\,1\,/\,3$$

So, in this case, the 2 feet 4-inch bar would weigh 29.8 ⅓

The next word problem that we will now move onto involves taxes, which makes it and its jargon a lot more useful to normal people than most of these other ones.

A property with an assigned value of $70,000 has a tax of $1,100. Meanwhile, there is another property within the same tax district as the original one. The tax imposed on this other property is $1,400. With this being mentioned, what is the assigned value of the second property mentioned here?

Here we are given two different categories to access: the assigned values of the property and their taxes. We should now put these into ratios to get our basic equation:

$$70,000\,/\,1,100 = v\,/\,1,400$$

The next step then would be to cross multiply. Our equation now looks something like this:

70,000 x 1,400 = v 1,100

Next, 98,000,000 = v 1,100

So, now we have to divide 98,000,000 by 1,100:

98,000,000 / 1,100 = 89,091

So, v here equals 89,091, and therefore the assigned value of the second property would be $89,091.

As you can plainly tell by now, ratios and proportions are extremely important to know how to multiply. This skill may even come in handy more often than does basic, single digit multiplication at times. If you stick to the principles laid out here, when solving for ratios and proportions, you will be able to get better with these problems and their equations.

## Chapter Three:
## Trigonometry and Its Uses in Real Life

Trigonometry is a subject that often gets a reputation as being harder than it truly is. To start off our discussion on trigonometry, this is the study of the calculations involving triangles in real life (hence the "trig" in trigonometry). As you can probably imagine, this subject is immensely useful as there are triangles everywhere you go in the world. Trigonometry, for the most part, works only with the lengths, heights, and angles of triangles. This field originated in the 3rd century BC and is now practiced by professionals such as crime scene investigators, engineers, physicists, astronauts, surveyors, and architects.

It may surprise you to learn that trigonometry actually stemmed more from astronomy than any field of mathematics, save geometry. And while trigonometry did not truly develop until around the third century BC, its origins date all the way back to around 2,000 BC.

Now that we have poured over some of the histories of trigonometry briefly, we should now look at some of the subject's everyday uses. Keep in mind that trigonometry's uses are wide and varied, so you cannot expect much depth in these topics for the sake of brevity.

### Physics
To start off with, trigonometry can be used in physics to perform a number of tasks. These include finding out the components of vectors that you come across, modeling waves and their mechanics ( both physical and electromagnetic waves), modeling oscillations of

waves, summing up the strength or weakness of fields ( of all sorts), and using dot and cross products. Another principle in physics that trigonometry can be applied to is projectile motion. It is understood that to someone who has no background in physics this all may seem like gibberish. If you want to find out more about these terms there are many great resources on the internet to do so.

## Navigation

The next practice that trigonometry can be used for is navigation. Trigonometry was more relevant in this respect to people living in earlier generations without mobile devices but if you are ever stuck without any means of navigation, some basic trigonometry could become useful for you to know.

The first and most important thing that trigonometry is used for in navigation is setting directions such as north, south, east, and west. With this, it can then tell you exactly which direction to take to achieve a straight line if you are using a compass. It can also be used to pinpoint certain locations, and, for mariners, it can be used for finding out a ship's distance from a shore as well.

## Criminology

It may come as a surprise to you to learn that trigonometry can also be used in criminology. The most important thing that trigonometry has to offer this field is its ability to calculate a projectile's trajectory through the air. Other important uses of trigonometry in criminology are the estimations it can provide as to might have caused a car crash or how an object fell from somewhere else. Trigonometry can also be used to determine at which angle a bullet has been shot. All of these uses can provide much-needed evidence for police and detectives and can also be used for virtually anyone who wishes to know details on cases.

## Marine biology

Another field in which trigonometry is often used that may also come as a surprise to you is marine biology. For example, one of the most common uses of trigonometry within this field is the acquisition of knowledge as to how a level of light affects the ability of algae to photosynthesize. Marine biologists are also known to use mathematical models to measure and analyze marine animals and their behaviors. In addition to these uses, marine biologists also use trigonometry to determine the sizes of larger animals from distances.

**Marine engineering**
In addition to marine biology, trigonometry is also very useful in marine engineering. In this field, trigonometry is often used to first build and then navigate marine vessels. Specifically, trigonometry can be used to create what is commonly known as the "marine arch", which is a sloping surface that connects higher level areas with lower ones. This forms a triangle, the components of which can be determined by trigonometry alone.

**Archaeology**
Another field in which trigonometry is commonly used is in archaeology. Archaeologists often use trigonometry to divide their excavation sites into smaller parts. They can also use trigonometry to determine their excavation procedures based on how far they need to dig down to find what they are looking for based on how old what they are looking for is. They also tend to use trigonometry to find out the distance they are from nearby sources of water.

While trigonometry has developed dramatically over the course of its lifetimes, there are a few principles from its inception which are still as valid as ever today. If you were to look back at virtually all of the notable inventions produced since the industrial revolution, you would be hard pressed to find one that does not owe, at least in some measure, its existence to trigonometry. One of the biggest breakthroughs in the history of trigonometry, however, was discovered by Galileo. Galileo discovered that any motion, whatever it may be of, consisted of two components acting on it: the vertical (or gravity) and the horizontal (or the object's projection). He also prognosticated that these two components should always be dealt with independently of one another. These findings caused scientists to gain the ability to measure the velocity of projectiles and the rate at which gravity would act against them, among other abilities.

**Surveying**
Another practice commonly used in trigonometry is what is known as surveying. One method of doing this is called triangulation which was first suggested by a mathematician by the name of Gemma Frisius. In this method, a person chooses a baseline of a known

length, and from this line's endpoints, the angles from it to more remote objects are then measured. This can be done with basic, elementary trigonometry. This process is usually repeated with multiple baselines until the entire area that is being studied is laid out in terms of triangles and their angles. It was a mathematician by the name of Willebrord Snell who first carried out this method on a large scale when he surveyed with 33 triangles an 80 mile stretch of land in Holland.

An even more ambitious survey was then carried out by a French astronaut by the name of Jean Picard, who triangulated the entire nation of France. After that, a yet more ambitious survey was carried out; the entire subcontinent of India was triangulated between the years of 1800 to 1913.

Now that we have gone over some of the everyday uses of trigonometry as well as some of its history, it would be beneficial to go over some of the most basic aspects of making trigonometric calculations.

**Trigonometric calculations**
The first point that should be touched on is the fact that trigonometry deals primarily with triangles that form right angles, with the sum of all of the internal angles totaling 90 degrees. Having a single right angle in a triangle makes it impossible for all of the sides to reach the same length. The angle opposite of the right angle is usually labeled as "O". The side opposite of the right angle is what is known as the hypotenuse. The opposite of the angle O, however, is known as the opposite. The other side next to the angle O besides the hypotenuse is known as the adjacent.

Next, we should go over the three main functions in trigonometry when it comes to triangles. Each of these is found by dividing the length of one side of a triangle by another. The first is called sine (or sin for short). To find the sin of a triangle, you must divide the length of the opposite by that of the hypotenuse. The next function is called cosine (or cos for short). To find the cos of a triangle, you must divide the length of the adjacent by that of the hypotenuse. The next and final function is called tangent (or tan for short). To find the tan, you must divide the length of the opposite by that of the adjacent.

## SOH CAH TOA

A commonly used abbreviation for finding the value of these functions out is what is known as SOH-CAH-TOA. Or, in equation form, Sin= opposite/ hypotenuse, Cos= adjacent/ hypotenuse, and finally, Tan= opposite/ adjacent.

Let's now look at a real-world example of how all of this could be used. Say you are building a roof and on either side of it the length of the opposite is 20 ft. and the length of the adjacent is 10 ft. Using these figures, we need to look for the hypotenuse's length. TOA will be the equation to use here. To find the tangent here we would take our opposite (10) divided by our adjacent (20). Here are equation would read Tan = 20/ 10, so, tan = 2.

While a calculator is sometimes needed for use on SOH CAH TOA problems with decimals, they can usually be solved by mental calculations with relative ease.

## Circles

Next, we should take a look at the uses of trigonometry when it comes to working with circles. The first thing that we should do is divide our circle into four quadrants. At the center of these four quadrants, what is known as a Cartesian coordinate in the center is 0,0. Any point left of the Cartesian coordinate has a negative x value. Any point below the Cartesian coordinate had a negative y value. So, a point in the upper left coordinate would be (-x, +y), a point in the upper right quadrant would be (+x, +y), a point in the lower left quadrant would be (-x, -y), and finally, a point in the lower right quadrant would be (+x, -y).

Now, if we were to determine a radius within this circle and we were to rotate that radius around the circle we would be left with a series of triangles. This is where our SOH CAH TOA equations come in handy yet again. If the radius was a length of 2 inches and it formed a right triangle with a duplication of the radius and the center of the circle, then the radius would be our adjacent and the duplication of the radius would be our hypotenuse. We would then need to find out the length of the opposite between the adjacent and the hypotenuse. This could be done with the equation CAH. Here we would plug in the adjacent (2 inches), and the hypotenuse (1 inch). So, we would then get 2/1= 2, so the length of the opposite side of the triangle would here be 2 inches.

We should now try out a word problem to get a better grasp on how to use trigonometry in everyday life. Let's say you are out sailing one day but you are not sure where you are going. You initially head out due east, and you do so at a cruising speed around 10 km/ hr. There is, however, also a tide here. It is due north at a speed of 5km/ hr. What direction are you going to end up traveling in under these circumstances?

To answer this question, you must first draw up your triangle. The length of this triangles adjacent would be 10 km, and the length of its opposite would be 5 km. Here we would need to use our tan equation to determine at which angle we are going to sail. This would give us here 5/10= .5.

The inverse tan of .5 is 26.6. This would, however, need to be subtracted from 90 as we are measuring it from 90%. So finally, we would be sailing 63.4% in the northeast direction.

Hopefully, none of these problems included in this chapter required a calculator for you. As you can plainly see by now, trigonometry has many more uses in everyday life than most people expect. If you stick to applying the principles laid out here, you are bound to come across scenarios in your everyday life in which you can apply them. All you have to do is look out for places in which these rules can be applied.

## Chapter Four:
## Adding and Subtracting Fractions

The next point that we are going to touch on here is adding and subtracting fractions. To not be intimidated if you are not a math person here, this chapter is going to be fairly straightforward and simple. It should also be noted that the material in this chapter is some of the most important and useful material that is going to be covered in this book. So, with that being said, it may be helpful to study this chapter in greater depth than you would some of the others.

To start off with, we should define "like fractions". "Like fractions" are fractions that have the same denominators. These fractions can be added and or subtracted with more ease than can other fractions. This is because to add or subtract these you only have to add or subtract their numerators. You would then just carry over the common denominator to your final answer.

If you want to add or subtract fractions with different denominators, you have to start off with finding equivalent fractions with the same denominators. There are two steps in doing this which are listed below:

1. Find out what the smallest multiple (LCM) is common to both of the numbers.
2. Next, write out the equivalent fractions with the LCM as the denominators in place of the original fractions.

When you are adding or subtracting fractions, the name for the LCM becomes the lowest common denominator (LCD).

As mentioned before, the addition or subtraction of fractions when their denominators are the same is the easy way. It only becomes difficult to do when denominators are different. Again, when adding or subtracting these fractions you must first find the LCM or, in other words, the LCD. Let's now look at an example of how to find these:

**Solving for fractions**
Take ¾. To find another way of writing this out we would need to find the lowest number divisible by both of these numbers. In other words, the smallest number that you can divide both of these numbers by. This number, in this case, would be 12. We would then need to multiply both of these numbers by 12 here. Our equation now would look something like this:

$$3 \times 12 / 4 \times 12$$

This would then equal:

$$36 / 48$$

In this example, 36 / 48 would be the equivalent fraction of 3 / 4. This is because both of the terms here have been multiplied by the same number, 12. So, to find equivalent fractions of fractions that do not have common denominators, you must first multiply them by the smallest number which they are both divisible by. Then you can get an equivalent fraction that will have larger terms but will be of the same value.

With that, we will move on to our slightly more complicated second example. This one includes the addition of fractions with different denominators.

$$¾ + ⅙$$

The first step we will need to take here is finding the lowest common denominator. This is done most easily by finding all of the multiples of each denominator and seeing which one is the smallest that the two sets have in common. We will now go over this method numerically:

4: 1 x 4= 4, 2 x 4=8, 3 x 4=12, 4 x 4=16

6: 1 x 6=6, 2 x 6=12. 3 x 6=18

As you can see, the first and lowest common multiple of the two is 12, so we would then use 12 as our common denominator.

The second method that we could use in this case is called prime factorization. We would do this by writing out each denominator as a product of its factors. The prime factors of 4 would be 2 and 2. As far as our common denominator is concerned, we need to use the factor which is included in both numbers. We would, therefore, need to use 2 twice and 3 once (3 because of 2 x 3= 6).

So, our prime factorization for 4 would be 2 x 2 and our prime factorization for 6 would be 2 x 3. Meanwhile, our LCD is 12.

Now that we have our lowest common denominator, we need to create equivalent fractions. We would now do this by multiplying each numerator and denominator by their respective factors. Since 3 x 4= 12, we would now multiply 3/ 4 by 3/ 3. Likewise, since 2 x 6 is 12, we would also multiply ⅙ by 2 / 2. These steps would then give us the equivalent fractions 9 / 12 and 2 / 12. At this point, we would need to add up our numerators, 9 + 2. Our answer then would be 11 / 12.

In equation form this would all look something like this:

*3 / 4 + 1 / 6 = 3 x 3 / 4 x 3 + 1 x 2 / 6 x 2 = 9 / 12 + 2 / 12 = 11 / 12*

Now we should take a look at another example to get a better grasp of what we are doing here.

Using the second method mentioned above, prime factorization, we will now solve for two fractions with mismatching denominators. These are 3 / 10 and 5 / 28.

Next, we would need to find the lowest common denominator.

2 x 5 = 10 and 2 x 2 x 7 = 28, both of these numbers have 2 in common. The number occurs once when multiplying to 10 and twice when multiplying to 28. Now we should take a look at all that we are multiplying and how many times these numbers all occur. Here we are multiplying 2 two times, 5 one time, and 7 one time, so our lowest common denominator would take all of these into account:

$$2 \times 2 \times 5 \times 7 = 140$$

Now that we have found our lowest common denominator to be 140 we need to divide this number by both of the original denominators. These equations would look something like this:

$$140 / 10 = 14, \; 140 / 28 = 5$$

Now we would need to create some equivalent fractions from the least common denominator for the two original fractions. Seeing as how the fraction 3 / 10 has a denominator of 10, we would have to multiply it by 14, in this case, to convert it to an equivalent fraction with the lowest common denominator of 140. Our equation for doing so would turn out to look something like this:

$$3 / 10 = 3 \times 14 / 10 \times 14 = 42 / 140$$

Seeing as how 5 / 28 have a denominator of 28, we would have to multiply this fraction by 5 to convert the denominator to those 140 marks. Our equation for doing so would look something like this:

$$5 / 28 = 5 \times 5 / 28 \times 5 = 25 / 140$$

Now that we have our lowest common denominator and we have adjusted our fractions accordingly, our final fractions would wind up looking something like this:

# MENTAL MATH

*42 / 140*

*and*

*25 / 140*

Let's now try the same thing with some different fractions. Let's now take 4 / 20 and 5 / 12 and use prime factorization to find their equivalent fractions. We should first start off by finding these two's lowest common denominator. We should now start this off by multiplying up each number's line until we arrive at the lowest common denominator:

*20 x 2 = 40, 20 x 3 = 60*

*12 x 2 = 24, 12 x 3 = 36, 12 x 4 = 48, 12 x 5 = 60*

As you can see here, our lowest common denominator, in this case, is 60. Seeing as how 20 x 3 = 60, we would need to multiply 4 x 3 next. This would equal 12. Seeing as how 12 x 5 = 60, we would then need to multiply 5 x 5. This would equal 25. So our final fractions after finding the lowest common denominator and factoring are 12 / 60 and 25 / 60.

Here it would be helpful to go over how to add and subtract mixed numbers. Mixed numbers are values that include both whole numbers and fractions in one. To start off adding or subtracting these, you must first add or subtract the whole numbers and then move on to the additional fractions. Let's start with some easier examples: 4 3/8 + 2 2/8. This one is easy because the fractions here have common denominators. First we would add up the whole numbers: 4 + 2 = 6, then we would add up the numerators: 3 + 2 = 5. The denominators do not need to be added up in this case because they are the same. Our final answer here would be 6 5/8.

Another example would be 3 2/5 + 1 4/5. Again, first, we need to add up our whole numbers and then add up our numerators. The denominators here do not need to be added up because they are the same. So first we would calculate 3 + 1 = 4. And then the numerators, 2 + 4 = 6. 6 / 5 are, however, improper. We would now need to convert this fraction into a mixed number in itself: 6/ 5 = 1 ⅕. So, 4 6/5 = 4 + 1 ⅕ = 5 ⅕. Our final answer here would be 5 ⅕.

Our next example is more complicated and it features denominators that are different from one another. The basic equation is as follows: 6 ¾ + 3 ⅝

Because these two denominators are different we need first to find the lowest common denominator here. Between 4 and 8 the lowest common denominator would be 8, so we would not need to multiply the numerator of the first fraction but we would need to multiply the first numerator, 3, by 2 because we had to multiply 4 by 2 to get to 8. So we would then come up with 6 6/8 + 3 ⅝. This would give us 9 11/8. 11/8, however, is an irregular fraction which we would need to convert to 1 ⅜. So now we are left with 9 + 1 ⅜ = 10 ⅜

Let's now look at another example of adding mixed numbers in which the denominators are different from one another, 8 2/4 + 6 ⅜.

The lowest common denominator here is 8. Since we would need to multiply 4 by 2 to arrive at 8, we would then need to multiply the first denominator, 2, by 2 as well. This would give us 4. Our new equation would then look like this: 8 4/8 + 6 ⅜. First, we would need to add up the whole numbers which would give us 14. Then, we would need to add up our numerators, which would give us 7. Our final answer here would be 14 ⅞.

As you can probably tell by now, adding and subtracting fractions and mixed numbers is not only easier than it may seem at first but it is also incredibly useful in everyday life. You deal with fractions all the time in life whether you are cooking from recipes, balancing your checkbook, or finding out novel statistics on the topics that interest you. You might as well learn how to deal with problems concerning fractions better when considering just how much of your time is spent dealing with these.

Mixed numbers are not all that complicated to deal with either and they prove to be very useful when dealing with difficult fractions that should not stand out alone by themselves. Upon investigating these you should find that mixed numbers are also very common and very easy to solve for. All of the principles laid out within this chapter should be feasible to follow mentally, tough if you feel that going over them while taking notes will in any way help you, you should try to do that.

THOMAS SCOFIELD

## Chapter Five:
# Mean, Median, and Standard Deviation in Everyday Life

Averages are extremely useful in that they tell us what the normative figures in a sequence are. There are many different types of averages, all of which give us different and unique perspectives on what is going on in the center of our distributions. Of these types of central measurement, the main ones are median, mean, average, mode, and range. This chapter will be dedicated to the study of these measurements.

Mode, median, and mean are all different types of averages. Averages are very common in any type of statistics but these three are usually the most common types that you will come into contact with. Of these three, the mean is probably the one which is most commonly calculated to by most people. To find the mean in a set of data, you must first add up all the numbers presented to you and then divide that sum by the amount of numbers in the set of data. The median is the middle of all of the numbers. This is usually much easier to find than the mean in any given data set. Just keep in mind that your numbers have to be arranged from smallest to largest when accounting for the median, so you may need to do some rearranging of figures before you are able to arrive at your number. The mode is the only one of these three that has a possibility of not occurring at all throughout a data set. The mode is simply the number that occurs most often throughout the set of data. If no numbers are repeated, or none are repeated any more than any others, then there will be no mode for the set of data. In addition, a range is a difference between the largest and smallest numbers within a set of data.

We should now start on some examples of value sets and how to solve for their averages. Let us now find the range, mode, median, and mean for this value set:

## Findings averages in data sets

*13, 18, 13, 14, 13, 16, 14, 21, 13*

We should start off by calculating the mean, as it is the most common average. We would do this here by adding up all of the terms and dividing by the number of all of the terms. This would look something like this in equation form:

*(13 + 18 + 13 + 14 + 13 + 16 + 14 + 21 + 13) ÷ 9 = 15*

So, our mean has turned out to be 15 in this case. You may have noticed that this number was not included in the original set of terms. This is common when finding the mean. They sometimes do not occur within the original data set.

Next, we need to find the median. The median is the middle value within the number set, so in order to find this, we first need to reorder our dataset from the lowest to the highest numbers. This would look something like this:

*13, 13, 13, 13, 14, 14, 16, 18, 21*

As you can see here, there are nine numbers included within this set. This means that our median would be found in the 5th number listed here because that is the halfway point between 9 and 1. In this case, the 5th number listed is 14, so our median here would be 14.

Next up we would need to find the mode. The mode represents the number which is most commonly repeated, so within this set-out mode would be 13.

And at long last, we come to our range. Again, this is calculated by subtracting the smallest number from the largest number. Within this set, 21 is our largest number while 13 is our smallest number. So our equation would read: 21 - 13 = 8. Our range, in this case, would be 8.

So, to recap, our mean here is 15, our median 14, our mode 13, and our range here is 8.

Let's now look at another example to get a better grasp on how to find these averages.

Let's take the number set: 1, 2, 4, and 7 for example.

First off, we would need to calculate the mean by adding up all the numbers within the set and then dividing then all by the number of terms, 4 in this case. This would look something like this in equation form: $(1 + 2 + 4 + 7) \div 4 = 14 \div 4 = 3.5$

As you can see, our mean here is 3.5 which is an irregular number. This is common when finding the mean.

Next, we need to find out what the median is. The median is the middle number within a data set. Here we do not need to reorder out numbers because they are already in numerical order. There is, however, no number in the middle here because this set has an even amount of terms. When this occurs we need to find out the halfway point between the two middle terms which are in this case 2 and 4. Our median would, therefore, be 3 in this case. 3 are, as you can see, not on this list at all, which will happen sometimes when you are finding the median.

Next, we need to find the mode. The mode is repeated oftentimes, however, none of the numbers listed in the set are ever repeated, so we would, in this case, have no mode.

And finally, we need to find the range of this set. This is done by subtracting the smallest number from the largest. Within this set, our smallest number is 1 while our largest number is 7. $7 - 1 = 6$, so our range, in this case, would come out to equal 6.

To recap, our mean for this data set is 3.5, our median here is 3, we would have no mode under these circumstances, and finally, our range here is 6.

This next example that we will now go over has many different characteristics than the ones that we have gone over previously. Let's now start with our standard set of numbers: 8, 9, 10, 10, 10, 11, 11, 11, 12, and 13.

To start off with, we first need to find the mean. This is, again, only done by adding up all of our terms within a set and then dividing the sum by the number of terms. In equation form this would all look something like this: (8 + 9 + 10 + 10 + 10 + 11 + 11 + 11 + 12 + 13) ÷ 10 = 105 ÷ 10 = 10.5

So, as you can see, our mean here is 10.5, which is an irregular number, but this is commonly the case with the mean.

Next, we need to find the median. The median, again, is the middle number within any given set of data. Here we have 10 terms in our set which is an even number. This means that there is no number in the middle of all of these. We would, therefore, need to find the value between the two innermost terms, 10 and 11. This would give us a median of 10.5 which is like our mean, an irregular number but this is common in finding the median.

Next, we would need to find the mode of this set. The mode is, again, the number which occurs most commonly within a set of data. Here we have two modes. These are 10 and 11 because they both occur three times each which is more common than any of the other terms listed here. Having two or more modes is another commonality in finding averages.

And finally, we come to find the range. We would need to find the range by subtracting the lowest term in the set from the largest term in the set. In equation form, this would look something like this: 13 - 8 = 5

So, as you can see, our range here would be 5.

To recap, our mean for this set turned out to be 10.5, our median is also shown here to be 10.5, we would have two modes in this example which would be 10 and 11, and our range is 5.

The mean and the median, in this case, turned out to be of the same values. This can occur often when finding out averages despite which averages the similarities happen between. When you find that two or

MENTAL MATH

more averages are of the same value within a set, do not take it as evidence that you did your calculations wrong as that will happen from time to time naturally.

Next, we should look at a word problem concerning averages. This will give us a better grasp of how to apply these concepts to real, everyday life.

The following are the test scores that a student got on the most recent tests: 87, 95, 76, and 88. He is hoping to get at least an 85 or better in the class. Considering these previous scores, what grade will he need to get on his final test at a minimum in order to meet his goal?

What we are trying to solve for here is the minimum grade that he needs to get in order to achieve this. First, we would need to find the average of his test scores by mean, or adding up all of his test scores and dividing the sum by the overall number of tests. Since the score for the last test is what we are trying to solve for here, this will be represented by x within the equation: $(87 + 95 + 76 + 88 + x) \div 5 = 85$

Once we have multiplied by 5, the equation then turns into this: $87 + 95 + 76 + 88 + x = 425$ which would in turn become this: $346 + x = 425$.

X would then equal 79, so he would need to get at least a 79% on his final test in order to achieve his goal of an average of 85% in the class overall.

Let's now take a look at another word problem.

A woman has four relatives. Their ages are 45, 36, 60, 18, and 20. What are the mean, median, mode, and range of this set of ages?

First, we need to calculate for the mean by adding up all of the numbers and then dividing the sum by the number of terms listed. In equation form this would look something like this: $(45 + 36 + 60 + 18 + 20) / 5 = 179 / 5 = 35.8$

As you can see, our mean here is irregular, which again is fine and normal.

Next, we would need to find our median. This is the number in the middle of the set, which in this case would be 36 numerically, so our median here is 36.

We would have no mode within this set because the mode is the most frequently occurring number within a set. None of the numbers listed above recur at all.

And finally, we need to find the range of this set. This is done by subtracting the smallest number from the largest number. As an equation, this would look something like this: 60 - 18 = 42. Our range, in this case, would, therefore, be 42.

As you can probably tell by now, finding out the values of the various types of averages within sets of data is a fairly straightforward process which does not provide very much difficulty once you initially learn how to perform all of these operations. Once you master these skills in making these calculations, you will invariably find that finding averages comes in handy all the time throughout your everyday life. There are averages within every set of numbers that you come across, so now that you have the tools to find these averages, you should use them to your full advantage when you find opportunities to do so.

# Chapter Six:
## Working with Conversion Factors

The next topic that we should go over is conversion factors. These are incredibly useful for anyone because we are in more or less constant contact with them every day. They come in all forms, some of which are easy to adjust to; others are more complicated and or vague in their conversions. Whatever way in which you find them, they come in handy when making everyday calculations nonetheless. We will now start out by going over some more esoteric conversion factors which most people do not come into contact with on a regular basis.

### Conversion factors
To convert acres to hectares you need to multiply by .4047. To convert acres to square feet, you need to multiply by 43,560. To convert acres to square miles, you need to multiply by .001562.

To convert atmospheres to centimeters of mercury, you need to multiply by 76. To convert Btu/hour to horsepower, you need to multiply by .0003930. To convert Btu to kilowatt-hour, you need to multiply by .0002931. To convert Btu/hour to watts, you need to multiply by .2931.

To convert bushels to cubic inches, you need to multiply by 2150. To convert 4 bushels (U.S.) to hectoliters, you need to multiply by .3524. To convert centimeters to inches, you need to multiply by .3937. To convert centimeters to feet, you need to multiply by .03281.

To convert cubic feet to cubic meters, you need to multiply by .0283. To convert cubic meters to cubic feet, you need to multiply by 35.3145. To convert cubic meters to cubic yards, you need to multiply by 1.3079. To convert cubic yards to cubic meters, you need to multiply by .7646.

To convert degrees to radians, you need to multiply by .01745. To convert dynes to grams, you need to multiply by .00102. To convert fathoms to feet, you need to multiply by 6. To convert ft. to meters, you need to multiply by .3048. To convert feet to miles (nautical), you need to multiply by .0001645. To convert feet to miles (statute), you need to multiply by .0001894. To convert feet/second to miles/hour, you need to multiply by .6818. To convert furlongs to feet, you need to multiply by 660.0. To convert furlongs to miles, you need to multiply by .125.

To convert gallons (U.S.) to liters, you need to multiply by 3.7853. To convert grains to grams, you need to multiply by .0648. To convert grams to grains, you need to multiply by 15.4324. To convert grams to ounces (avdp), you need to multiply by .0353. To convert grams to pounds, you need to multiply by .002205.

To convert hectares to acres, you need to multiply by 2.4710. To convert hectoliters to bushels (U.S.), you need to multiply by 2.8378. To convert horsepower to watts, you need to multiply by 745.7. To convert horsepower to Btu/hour, you need to multiply by 2,547.

To convert hours to days, you need to multiply by .04167. To convert inches to millimeters, you need to multiply by 25.4000. To convert inches to centimeters, you need to multiply by 2.5400. To convert kilograms to pounds (avdp or troy), you need to multiply by 2.2046. To convert km. to mi., you need to multiply by .6214.

To convert kilowatt-hour to Btu, you need to multiply by 3412. To convert knots to nautical miles/hour, you need to multiply by 1. To convert knots to statute miles/hour, you need to multiply by 1.151.

To convert liters to gallons (U.S.), you need to multiply by .2642. To convert liters to pecks, you need to multiply by .1135. To convert liters to pints (dry), you need to multiply by 1.8162. To convert liters to pints (liquid), you need to multiply by 2.1134. To convert liters to quarts (dry), you need to multiply by .9081. To convert liters to quarts (liquid), you need to multiply by 1.0567.

To convert meters to feet, you need to multiply by 3.2808. To convert meters to miles, you need to multiply by .0006214. To convert meters to yards, you need to multiply by 1.0936. To convert metric tons to tons (long), you need to multiply by .9842. To convert metric tons to tons (short), you need to multiply by 1.1023. To convert miles to kilometers, you need to multiply by 1.6093. To convert miles to feet, you need to multiply by 5280.

To convert miles (nautical) to miles (statute), you need to multiply by 1.1516. To convert miles (statute) to miles (nautical), you need to multiply by .8684. To convert miles/hour to feet/minute, you need to multiply by 88.

To convert millimeters to inches, you need to multiply by .0394. To convert ounces (avdp) to grams, you need to multiply by 28.3495. To convert ounces to pounds, you need to multiply by .0625. To convert ounces (troy) to ounces (avdp), you need to multiply by 1.09714. To convert pecks to liters, you need to multiply by 8.8096. To convert pints (dry) to liters, you need to multiply by .5506. To convert pints (liquid) to liters, you need to multiply by .4732.

To convert pounds (ap or troy) to kilograms, you need to multiply by .3732. To convert pounds (avdp) to kilograms, you need to multiply by .4536. To convert pounds to ounces, you need to multiply by 16. To convert quarts (dry) to liters, you need to multiply by 1.1012. To convert quarts (liquid) to liters, you need to multiply by .9463.

To convert radians to degrees, you need to multiply by 57.30. To convert rods to meters, you need to multiply by 5.029. To convert rods to feet, you need to multiply by 16.5. To convert square feet to square meters, you need to multiply by .0929. To convert square kilometers to square miles, you need to multiply by .3861. To convert square meters to square feet, you need to multiply by 10.7639. To convert square meters to square yards, you need to multiply by 1.1960. To convert square miles to square kilometers, you need to multiply by 2.5900. To convert square yards to square meters, you need to multiply by .8361.

To convert tons (long) to metric tons, you need to multiply by 1.016. To convert tons (short) to metric tons, you need to multiply by .9072. To convert tons (long) to pounds, you need to multiply by 2240. To convert tons (short) to pounds, you need to multiply by 2000.

To convert watts to Btu/hour, you need to multiply by 3.4121. To convert watts to horsepower, you need to multiply by .001341. To convert yards to meters, you need to multiply by .9144. To convert yards to miles, you need to multiply by .0005682.

As you can probably already tell, there are no great methods of memorizing these. You can only master them by rote memorization which is not very efficient but will still give you the greatest results possible in this case.

When thinking about conversion factors it is useful to see them as a means of changing the units of a measured quantity without changing its value. There is one method of calculating the value of a conversion factor. This is called the unity bracket method. The unity bracket method involves setting the numerator and denominator of a fraction at the same value in order to determine the applicable unit conversion factor.

Conversion factors are, as you can probably already tell, very easy to deal with, they just take time to memorize. Once these are memorized, however, they do very much come in handy in everyday life and are usually easily committed to long-term memory. The only hard part of dealing with these factors is the initial memorization of certain multiplications that is necessary. Keep in mind also that you do not need to memorize all of the conversion factors mentioned above. You would actually be wiser to only memorize the ones that you expect are going to be immediately useful to you in your everyday life.

# Chapter Seven:
## Using the PIN Technique

Next, it would be helpful for us to go over the plugging in numbers or PIN technique. This technique is especially helpful for students because it offers a fast and easy way to make calculations without any calculator. On the ACT and the SAT tests, as well as any other standardized tests which include math, the only thing that matters to your graders is that you arrived at the correct answer. It is not relevant by what means you came to the answer. This gives you much-needed flexibility as you do not necessarily have to stick with the methods taught to you by your teachers in order to pass these tests. Below we are going to go over why and how to use the PIN method when making your calculations.

First, we should start out with why this method is useful. There are bound to be a lot of problems that you simply do not know how to approach, have too many variables for you to know their right answer, or seem like they would take too long for you to solve. These circumstances are where the PIN method can prove to be rather useful.

Let's first take solving problems with lots of variables for example. This can seem intimidating and overly time consuming at first but if you replace all of the variables within these problems with certain numbers it can make the whole process a lot more bearable and a lot less time-consuming. It will also make these problems look a lot more accessible and a lot less confusing. Let's now look at an example of one of these problems in which we would apply the PIN method:

## The PIN method

For all numbers that represent a and b, let a b be defined as a b = ab + a + b. For all numbers that represent x, y, and z, which of the following is true?

$$1.\ x\,y = y\,x$$

$$2.\ (x-1)(x+1) = (x\,x) - 1$$

$$3.\ x(y+z) = (x\,y) + (x\,z)$$

Answers: a) 1 only, b) 2 only, c) 3 only, d) 1 and 2 only, e) 1, 2, and 3.

This answer will be brought up below.

It can be easy to forget that you have this method at your disposal when you are taking a test. Keep in mind though that this method will help you to simplify the problems that you come across greatly. Whenever you see a problem on a test that includes more variables than you are comfortable with, you should keep this method in mind for such occasions.

Now that we have covered why the PIN method is helpful, we should delve into how to go about using it. The basic idea of the PIN method, as mentioned before, is that you replace the variables that you are faced with other numbers. The technique can work for any problem in which you are confronted with variables whether it is geometry or algebra that you are doing.

A general rule of thumb for using the PIN method is that you can use it whenever there are variables in the equation. The more variables that an equation has, the more useful the PIN method becomes.

Because these problems involving variables determine relationships between numbers, you can tell within these which relationships are constant and which are not. In other words, you can see which relationships hold true regardless of the numbers being used. As long as your own numbers follow the rules set forth within the equation, you can use your own numbers in replacement of the variables originally set forth and arrive at the correct answer.

Once you have picked your desired number to represent the variable, you can then solve the original equation with that number. After that, you can then look for the original variable in your answer options and replace it with the new number that you have chosen. By doing this you can more easily test your options as far as answers go and see which of the answers best matches the results you came up with for the problem. If this does not make total sense yet, do not worry. We are now going to go over the previous problem in more detail to get a better grasp of what we are doing here.

For all numbers that represent a and b, let a b be defined as a b = ab + a + b. For all numbers that represent x, y, and z, which of the following is true?

1. $x\,y = y\,x$

2. $(x-1)(x+1) = (x\,x) - 1$

3. $x(y+z) = (x\,y) + (x\,z)$

Answers: a) 1 only, b) 2 only, c) 3 only, d) 1 and 2 only, e) 1, 2, and 3.

We are told that the relationship mentioned above applies equally to all of the numbers x, y, and z. We are therefore able to replace any of these letters x, y, and z with any given numbers here because any and all numbers would work in this case.

We should now give each of these variables their own numbers. Seeing as how they are all different, they should each get their own individual numbers assigned to them. This would look like this:

$$x = 2,\ y = 3,\ z = 4$$

Let's now solve this problem using the new numbers that we have assigned to these variables and see if everything works out the same for us using this method.

The first part is:

$$x\ (+)\ y = y\ (+)\ x$$

We should now take half of this equation and replace the variables with our new numbers: 2 (+) 3

This would in turn become:

$$(2)\ (3) + 2 + 3 = 11$$

Now that we see that the left half of our equation equals 11 we should solve for our right half:

$$y\ (+)\ x = 3\ (+)\ 2 = (3)\ (2) + 3 + 2 = 11$$

So, as you can now see, both of our halves equal one another in this case. Now we will refer back to our answers:

$$1.\ x\,y = y\,x$$

$$2.\ (x-1)\,(x+1) = (x\,x) - 1$$

$$3.\ x\,(y+z) = (x\,y) + (x\,z)$$

Answers: a) 1 only, b) 2 only, c) 3 only, d) 1 and 2 only, e) 1, 2, and 3.

Now that we know that both sides equal 11, we can deduce from that that the first statement is true. This, in turn, means that we can disregard b and c as possible answers because they exclude the first statement listed here.

We should now try the PIN method on statement to see if it is also true. Let's first take a look at the original equation:

$$(x-1)\,(x+1) = (x\,x) - 1$$

Again, we should start by applying our numbers to the left side of the equation first. This would look something like this:

$(x-1)(+)(x+1) = (2-1)(+)(2+1) = 1(+)3 = (1)(3)+1+3 = 7$

So now that we know that the left side of out equation equals 7, we should move on to the right side of the equation. The original equation on the right side looks like this:

$$((x(+)x))-1$$

Now we would have to replace the xs here with our replacement number, 2.

$$((2(+)2))-1 = ((2)(2)+2+2)-1 = 7$$

So here our left side equals 7, as does our right side as well. This means that statement number 2 is also correct. We can then eliminate answer a in addition to answers b and c.

Our final step would be to replace the variables in the last statement with the numbers that we have decided on. The original equation here looks something like this:

$$x(y+z) = (xy)+(xz)$$

First, we need to plug in our substitute numbers into the left side of the equation. After doing so the left side equation turns into this:

$$2(+)(3+4)$$

Which would then become:

$$2(+)7 = (2)(7)+2+7 = 23$$

So, as you can see, the left side of this equation equals 23. Next, we need to solve for the right side in order to see if the result matches the answer for the left side. The original equation for the right side looks something like this: $(x\,y)+(x\,z)$.

After substituting the variables with our new numbers we would come up with this:

$$((2\,(+)\,3)) + ((2)\,(+)\,4)) = ((2)\,(3) + 2 + 3) + ((2)\,(4) + 2 + 4) = (7) + (14) = 25$$

So our answer for the right side of the equation here is 25. Our answer for the left side of this equation was, again, 23, which means that the two sides are not equal and therefore statement number 3 is not correct in this case. This would leave us with answer d as the only correct answer. Statements 1 and 2, in this case, are correct statements, while statement 3 is incorrect.

We were able to choose all of our own numbers in this example, but do keep in mind that this is not always the case when using the PIN technique. Always be on the lookout for when you can choose your own numbers for most, if not all, of the variables in a given equation or you can choose your own number for one of the variables and solve for the rest. This will give you a very helpful advantage in solving problems with more efficiency. We were able to choose our own numbers for all of the variables in the problem mentioned above only because we were told that the variables applied to all numbers in that case. Any number that we could have chosen would have followed the rules laid out for us.

You should only assume that you are able to plug in your own numbers in place of variables when the problem tells you specifically that the variables apply to any and all numbers. Usually, when you do not see this directive written out explicitly, you are allowed to choose your own number for one variable while still solving for all or the others. This will ensure that all of the variables are following their own rules and are keeping the relationships between one another intact.

We should now solve for an example of a problem in which we do not have the option of determining all of our own variables.

$$x = 3v,\ v = 4t,\ x = pt$$

Given the set of equations listed above, if $x = 0$, what is p's value?

We are not told here that this problem applies to any and all numbers which are why we can only make our substitution for one variable and leave the rest alone. To solve here we are going to replace v with

# MENTAL MATH

our own number. This is because v shows up in the middle equation, which shares the variables of the other two equations.

The first thing that we should note now is the fact that v = 4t. With this in mind, it becomes clear that we should choose a number divisible by 4 to take the place of v., In this case, let's now say that v = 8. So now if we replace every v in every equation our first equation will then look like this:

$$x = 3\,(8) = x = 24$$

So, as you can now see, when v = 8, x = 24 in turn. Now we need to move on to our next equation. The original version looks like this: v = 4t

When substituting for v we would then get

$$8 = 4t = t = 2$$

So, when x = 24 and v = 8, t = 2 in turn.

Our final step would be to now take a look at the last equation using our newfound values for the variables. The original equation looks like this:

$$x = pt$$

When the variables are substituted this equation turns into this:

$$24 = p(\,2\,) = p = 12$$

So, as you can see, p would equal 12 in this case. You could now make the assertion that p may not equal 12 in this case if v had not been originally determined to equal 8. We could then test this assertion by given v a different value. Let's now say that v = 20.

Our first equation under these circumstances would turn into this:

$$x = 3\,(20) = x = 60$$

Our second equation would then turn into this:

$$20 = 4t = t = 5$$

And our final equation would then turn into this:

$$60 = p\,(5) = p = 12$$

So, as you can see, p again equals 12 here. This is because we have kept our variables intact while solving both times. It is not relevant what any of the other variables are determined to be so long as we keep these variables intact. Our final answer, in this case, would be 12, or:

$$p = 12$$

So those are the bigger points on using the PIN strategy. Now that you have some of the skills concerning the use of this strategy, it should be easy to apply these when you are taking any tests with problems that may require this strategy in the future. The PIN strategy is also helpful in everyday life as well. It will come in handy whenever you have problems with too many variables for you to be comfortable with. Keeping this strategy in mind in your future as well as its application can give you a great advantage in math and general problem-solving.

## Chapter Eight:
# Techniques for SAT, GMAT, and GRE Students

This next chapter will be devoted exclusively for use by SAT, GMAT, and GRE students. We will focus mainly on SAT practice questions here because these are similar enough to the questions of the other tests to still be useful for students preparing for the GMAT or the GRE. Feel free to skip over this chapter if you are not studying for any of these tests and feel that this will not be a great use of your time.

To that end, our first practice question is an algebraic word problem.

**SAT problems**
Within a classroom at central high school, the mean number of students (or y) can be determined by the equation $y = 0.8636x + 27.227$. X, in this case, represents the number of years it has been from 2004, it is less than or equal to 10. Of the following statements, which is the best interpretation of the significance of the number 0.8636 within the context of this classroom?

    a.    The mean number of students in the classroom in the year 2004.
    b.    The mean number of students in the classroom in the year 2014.
    c.    The yearly decrease in the mean number of students in the classroom.
    d.    The yearly increase in the mean number of students in the classroom.

To answer this question, we would need to determine the slope of the equation and its relationships to the real world situation it models. You should also keep in mind that we are only solving for the independent variable, y, here instead of the dependent variable, x.

Choice d here would be the correct one. Let's now determine why this is. When this equation is written in y= mx+ b, or slope form, the coefficient of x (or 0.8636 in this case) would be what is known as the slope. The slope of this equation would then give you the amount by which the mean number of students in the classroom changes on average each year. The slope here would be a positive one which means that there is an increase in the mean number of students each year.

Let's now move on to our next example:

If $2/a - 1 = 4/y$, and $y \neq 0$ where $a \neq 1$, what is y in terms of a?

  a)  y = 2a − 2
  b)  y = 2a − 4
  c)  y = 2a − 1 / 2
  d)  y = 1 / 2 a + 1

Choice a is the correct one in this case. Now we will go over why it is.

First, you need to cross multiply the denominators by their opposite numerators which would be 2 x y = 2y, and 4 x (a - 1) = 4a - 4.

So, as you can see, this would give you the equation

$$2y = 4a - 4$$

To solve this you would then divide both sides by 2 in order to isolate the variable y.

So, y = 2a - 2 in this case. Therefore, the correct answer would be a here.

Reliable SAT, GMAT or GRE study would have to go very far beyond the problems mentioned here in order to be effective. These are very long and expansive tests that need to be taken seriously with long hours of devoted and focused study beforehand by those who take them. There are innumerable resources on preparing for these tests, so if you have one of these coming up it would be very beneficial for you to look into more on this subject. It should also be noted that we have only covered some of the algebra questions featured on the test here. The actual test itself goes over many other types of math than just algebra.

These tests also involve sections in which students are tested upon other subjects, so if you are about to take one of these tests, make sure that you are studying an array of subject matter in order to be prepared for all dimensions of the test at hand.

THOMAS SCOFIELD

## Chapter Nine:
## Math Strategies for Anyone

This book is, however, not just meant for use by students preparing for exams. This book is written for anyone who desires better mental math skills. Now we should go over some of the most basic aspects of mental math that you should keep at your disposal at all times in order to be better able at making calculations on the fly.

One aspect of mathematics that any adult can appreciate and apply to his or her everyday life is adjusted gross income. This is defined as the sum of the income that a person acquires within a year. Let's now look at an example of how to solve for adjusted gross income.

**Adjusted gross income**
In 2009, a man named Peter made a total of $15,000 working at his main job. In addition, he also made 200 extra dollars working another job for 20 weeks. Using these figures, we now need to find Peter's adjusted gross income.

*Earnings = weekly earning × number of weeks worked*

*Earnings = 200 × 20 = 4,000*

*Adjusted gross income = wages + interest income*

*Adjusted gross income = 15,000 + 4,000 + 50*

Peter's adjusted gross income would, therefore, be = 19,050 USD.

In 2009, a woman named Martha earned a total of $80,000 at her primary job. In addition, she also earned $50,000, $20,000, and $100,000 from her side jobs. Using these figures, we now need to determine Martha's adjusted gross income.

What is Martha's adjusted gross income?

*Adjusted gross income = wages + interest income*

*Adjusted gross income = 80,000 + 50,000 + 100,000 + 20,000*

Martha's adjusted gross income would, therefore, be = 250,000 USD.

Next, we should take a look at how to calculate for price discounts. Price discounts are when the price of an item is marked down, usually by a certain percentage. Here we will go over how to determine the new price of an item based off of the list price and the discount rate of the item.

**Price discounts**
Here is our first example: we need to find the new price of an item with a list price of 100 dollars and a discount rate of 25%.

*Discount = 100 × 25% = 100 × 0.25 = 25*

*Sale price = List Price - Discount Price = 100 - 25 = 75 dollars*

And on to our second example: here we need to find the sale price of an item that has a list price is 24 dollars and a discount rate of 50%.

*Discount = 24 × 50% = 24 × 0.50 = 12*

*Sale Price = List Price - Discount = 24 − 12 = 12 dollars*

Now we should go over some math for investment strategies. Here we will delve into calculating for simple interest. In its simplest form, the equation for simple interest is as follows: Interest = Principal × Rate of Interest × Time

## Simple interest

We should now look at an example of how to apply this equation. Here we need to compute the interest if the principal is 2,000,000 dollars at a rate of interest of 4% for a year.

We may need to use a calculator here:

$$Interest = 2{,}000{,}000 \times 4\% \times 1$$

$$Interest = 2{,}000{,}000 \times 0.04 \times 1$$

$$Interest = 80{,}000 \times 1 = 80{,}000$$

And here is another example of calculating simple interest in real life. Here we need to compute the interest if the principal is 100 dollars at a rate of interest of 2% for 10 years.

Using a calculator:

$$Interest = 100 \times 2\% \times 10$$

$$Interest = 100 \times 0.02 \times 10$$

$$Interest = 2 \times 10 = 20$$

Next, we will go over mortgage loans. This advice will be particularly useful for anyone considering buying a home.

## Mortgage loans

In this example, you are buying a house for $250,000. Towards this, you initially make a down payment of 15% of the purchase price and then you agree to a 30-year mortgage to cover the balance.

With all of this in mind, what is your down payment? And what is your mortgage?

$$Down\ payment = Purchase\ Price \times Percent\ Down$$

$$Down\ payment = 250{,}000 \times 0.15 = 37{,}500$$

*Amount of Mortgage = Purchase Price − Down Payment*

*Amount of Mortgage = 250,000 − 37,500 = 212,500*

If let's say, your monthly payment is 1,200 dollars, what is the total interest charged over the life of the loan?

*Total Monthly Payment = Monthly payment × 12 Months per year × Number of years*

*Total Monthly Payment = 1,200 × 12 × 30 = 432,000*

*Total Interest Paid = Total Monthly Payment − Amount of Mortgage*

*Total Interest Paid = 432,000 − 212,500 = 219,500*

This concludes our consumer math section of the book. The section should be immediately useful for most people than are any others. If you apply the equations mentioned here the next time that you find yourself dealing with mortgage loans, simple interest, discount prices, or adjusted gross income, you will be better able to make decisions regarding these matters.

# Conclusion

Thank you for making it through to the end of *Mental Math: Tricks and Practical Strategies to Make Calculations Faster, Enhance Your Math Skills and Solve Everyday Math Problems Easily*. Let's hope that this book was as helpful and informative as possible for you. Mental math is always a great skill to have no matter what profession you have or what your greater interests are in life. Developing skills in mental math will be very beneficial in your everyday functioning. Do not give yourself the impression that there is no room to improve on your mental math skills from here. The next step in your education beyond finishing this book would be to look into other materials on this subject. There are a lot out there and they will all get you even closer to becoming a true math whiz.

Mental math skills are a cumulative process. If you truly want to get better at performing calculations in your head then you are going to need to be diligent and persistent in your study of these skills. Again, thank you for purchasing this book. Let's hope it was as helpful and informative as possible.

## Other Books By Thomas Scofield

Discover 21 Powerful Techniques You Can Use To Build Self-Discipline And Achieve Your Goals Faster.

Do you want to become more productive? Do you want to learn how to achieve your goals faster? Have you ever looked at successful people in your life and asked what was their secret? If you said yes to any of these questions then you'll love this book.

One of the best ways to improve your productivity and achieve your goals is learning how to build self-discipline. When you know how to convert bad habits into productive ones it will be a lot easier to reach your goals and accomplish more in life.

However, becoming disciplined may be hard if you do it the wrong way. Avoiding temptations and procrastination everyday can't really be done without learning how to build powerful self-discipline habits.

In this book you'll find 21 proven techniques that will help you convert bad habits into good ones, skyrocket your productivity and achieve your goals faster than ever before. This isn't a book full of theoretical fluff. You'll learn practical techniques that you can actually put to use right away and that will help you accomplish things more easily thanks to the power of self-discipline.

In this book you'll discover:
- 21 Proven Techniques To Develop Powerful Self-Discipline Habits And Achieve Your Goals Faster
- The Right Way To Use Incentives To Become More Self Disciplined
- What Are The 2 Types Of Discipline And How They Can Improve Your Life
- How To Organize Your Life And Keep Things Simple
- Practical Tips To Sleep Enough And Still Have Time To Accomplish Your Goals
- How To Choose A Mentor To Help You Become Disciplined And Achieve What You Want In Life
- How To Better Manage Your Time To Be More Productive
- How To Teach Your Mind To Become Disciplined Using Physical Exercise
- Why Removing Temptations From Your Life Will Empower You
- 9 Affirmations You Can Use To Grow Into A More Disciplined Person
- Tips And Tricks To Get Your Diet On The Right Track
- When And How To Say "No" To People
- And Much, Much More

**Learn how to become more disciplined and achieve your goals!**

**"How To Build Self Discipline" by Thomas Scofield is available at Amazon.**

Discover How To Read Faster, Improve Your Memory And Learn Any Subject In A Short Period Of Time.

The pace of life is accelerating, knowledge is constantly growing and becoming more accessible. In today's society work and school are becoming more competitive, and if you want to stay ahead, you're constantly expected to know more and more and act faster and faster. Our time however, is still the same, so how can you keep up?

Accelerated Learning may be the solution for you, because it will help you acquire knowledge and new techniques at an accelerated speed, saving you time and money and giving you an edge over your competition.

In this book you'll discover how to improve your reading speed, develop your memory, acquire new skills faster and quickly learn any subject following the Accelerated Learning strategies. Whether you're a student looking to make the most of your time, career professional looking to acquire new skills to land your dream job, teacher or employer wanting to provide job training, this book will help you develop your learning ability and reach your goals faster.

In this book you'll discover:
- How To Learn Any Subject Faster Following The 5 Phases Of Accelerated Learning
- The Benefits And Outcomes Of Accelerated Learning
- The Theory Of Learning And How It Affects Your Performances
- How To Improve Your Memory Through Repetition, Organization And Elaboration
- A 3-Step Process To Quickly Understand Any Text
- 4 Simple Techniques To Improve Your Reading Speed
- How To Deeply Understand A Text Following The Socratic Method
- 6 Powerful Tools To Accelerate Your Learning Process
- How Organizing Your Space And Time Can Improve Your Memory And Help You Learn Faster
- Complete Lists Of Additional Books And Resources On Accelerated Learning
- And Much, Much More

**Discover the secrets to learn any subject faster and achieve your goals!**

**"Accelerated Learning" by Thomas Scofield is available at Amazon.**

www.ingramcontent.com/pod-product-compliance
Lightning Source LLC
Chambersburg PA
CBHW071429220526
45469CB00004B/1464